A PLAYFUL MATH SI[illegible]

WORD PROBLEMS *from* *Literature* STUDENT WORKBOOK

SECOND EDITION

DENISE GASKINS

Tabletop Academy Press

Tabletop Academy Press, Boody, IL, USA
tabletopacademy.net

ISBN: 978-1-892083-67-8
Print version 2.0

About the Font:
The Cadman font was designed by P.J. Miller designed to be as reader-friendly as possible. Letters are distinct and easy to distinguish, especially those most easily confused by students and people with dyslexia. Cadman is suitable wherever a clear and legible sans serif font with open type features is required.

Contents

For answers and fully worked-out solutions, see the companion book *Word Problems from Literature: Help Students Master Problem Solving in Elementary to Middle School Math.*

Storying—
encountering the world
and understanding it contextually
by shaping ideas,
facts,
experience itself
into stories—
is one of the most fundamental means
of making meaning.
As such, it is an activity
that pervades all learning.

—GORDON WELLS

Math ain't about numbers.
If you think that math is about numbers,
you probably think that Shakespeare
is all about words.
You probably think that dancing
is all about shoes.
You probably think that music
is all about notes.

Math ain't about numbers!
Math is about logic,
it's about beauty,
it's about connections.
It's about how you get
from one place to another.

—Cliff Stoll

1

Tools for Mathematical Problem-Solving

ARCHIMEDES TRIED TO FIND THE distance around a circle and almost discovered calculus. Pierre de Fermat predicted the result of a gambling game and laid the foundations of probability. Leonhard Euler went for an afternoon walk over the bridges of Königsberg and invented topology. Georg Cantor created a way to count infinity and opened a whole new world of modern math.

Through the centuries, mathematics has grown as mathematicians struggled with and solved challenging puzzles.

Problems are the raw material of math, the ore we dig, grind up, and melt, refining it to produce ideas. Our understanding of math grows as we play with problems, puzzle them out, and look for connections to other situations. The threads that connect these problems become the web of ideas we call mathematics. Each puzzle we solve adds a new thread to the web, or strengthens one that already exists, or both.

What If I'm Not Good at Math?

Do you struggle with math problems? Many people, when they get stuck or frustrated, decide that they're just not good at math.

But the truth is that *nobody* is good at math, if you define "good at math" to mean they can see the answer instantly. Here's a more useful definition: You're good at math if you have problem-solving tools and know how to use them.

And *that* is something everyone can learn.

When you're faced with a math problem, you need to combine the given facts in some way to reach the required answer. But rarely can you do it in a single leap. So you need to take one little step at a time. Can you think of a way to get closer to the goal?

Ask yourself these four problem-solving questions...

(1) What Do I Know?

Read the problem carefully. Reread it until you can describe the situation in your own words.

List the facts or information given in the problem. Notice math vocabulary words like *factor, multiple, area,* or *perimeter.* What do you remember about those topics?

Sometimes a problem tries to trick you. Watch out for mixed units: If one length is given in inches and another is given in yards, make them consistent.

Try to express the facts in math symbols or using the visual algebra of a bar model diagram.

To solve math problems, be like a detective looking for clues.

(2) What Do I Want?

Describe the goal, what the problem is asking you to find. What will your answer look like?

Notice important words like *product, sum, next,* or *not.* Small words like "not" are especially easy to miss.

Try to express the goal in math symbols or using the visual algebra of a bar model diagram.

(3) What Can I Do?

Imagine yourself in the story situation, applying your hard-earned common sense. If this actually happened to you, what would you do?

Mix things around in your mind. Combine the given facts. Have you worked a problem like this before? How did you solve that one? Will that method, or something like it, work here?

If you're using a bar model, think about ways you might move or cut the blocks to discover new relationships.

Try a tool from your problem-solving toolbox:

- Draw and label a diagram or sketch.
- Act the problem out, step by step.
- Make a systematic list, chart, table, or graph.
- Look for a pattern.
- Simplify the problem. Try it with smaller numbers.
- Change your focus. Restate the problem in another way.
- Look for a related problem. How is it the same? How is it different?
- Think about "before" and "after" situations.
- Work backwards. Start at the end of the problem and find a path back to the beginning.

- Guess and check. Try something to see if it works, and then make sure you know why.

If you are completely stumped, explain the problem to another person. Talking it over might help unclog your brain, opening your eyes to clues you had missed until you put them into words. Or take a break to let the problem simmer in your unconscious mind while you do other things.

(4) Does It Make Sense?

Don't neglect this last step! When you think you have found the answer, always look back at the original problem one more time.

Does your answer make sense? Did you leave anything out?

Can you think of a way to confirm that your answer is right?

Can you think of another way you might have gotten the answer? If you see an alternate approach, would that method have been easier? Make a note of any ideas you come up with. You may need them to solve your next puzzle.

What Are Bar Model Diagrams?

Bar model diagrams are algebra in pictures. We show all the known and unknown parts of our story as block-like rectangles. Then we imagine moving the blocks around or cutting them into smaller pieces until we can figure out the unknown parts.

Bar models transform the abstract mystery of the word problem into a shape puzzle: How can we fit these blocks together? This helps us notice the underlying structure of a word problem, so we can see how it's like others we've already solved, just as detectives search for a criminal's *modus operandi* (MO). Recognizing a math problem's structure helps us solve the case.

It's almost like magic. A trick well worth learning, no matter what math program you use.

Basic Bar Models

All bar model diagrams descend from one basic principle: The whole is the sum of its parts. If you know the value of both parts, you can add them to get the whole. If you know the whole total and one part, you can subtract what you know to find the other part.

Whole

Part A | Part B

The most basic bar diagram: Two parts make a whole.

Suppose we had a problem like this: “How much must I add to 2 to get 7 as the sum?” Would the words *add* and *sum* trick you into saying 2 + 7 = 9?

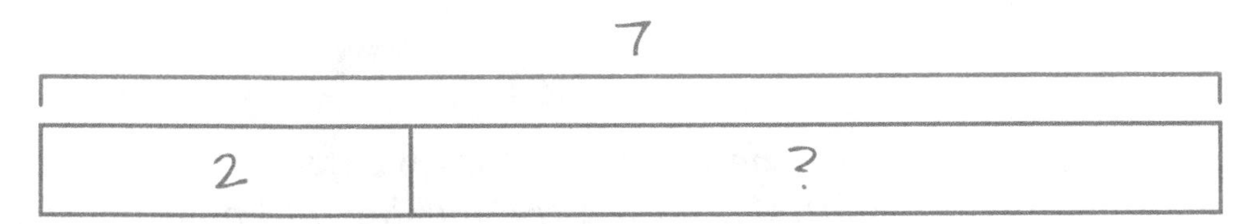

Seven is the sum, the whole thing.
What must be the missing part?

We can draw a rectangular bar to represent the total amount. Then we divide it into two parts, representing the number we know and the unknown part. Now it’s easy to see the answer.

Drawing Tougher Puzzles

As the number relationships in math problems grow more complex, we can split our bar model into more than two parts. Also, the relationships in our story may require a more elaborate diagram.

Multiplication, division, or fraction problems will involve several parts that are the same size, called units.

But even with a complicated story, the solution begins by drawing a simple bar to represent one whole thing.

For instance: "There are 21 girls in a class. There are 3 times as many girls as boys. How many boys are in the class?" Many students will multiply 21 × 3 to get the answer. But a bar model will keep us from falling into that trap.

To show three times as many girls as boys, we start with a bar for the number of boys. That's one unit, and then we need three units to show the number of girls.

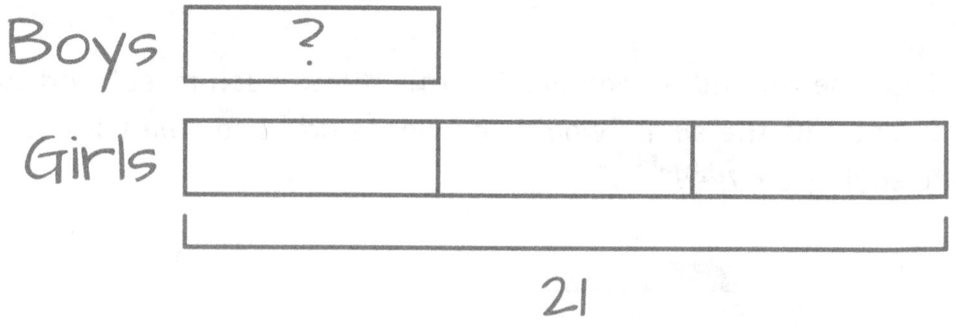

Boys ?

Girls

21

The bar model helps us see that we need to divide, not multiply, to find the number of boys.

Master These Two Rules

When you want to draw a bar diagram, start with long rectangles—imagine Lego blocks or Cuisenaire rods. Think of the bar as all of the books, fish, snowballs, or whatever from your story set out in a row.

Write labels beside each bar, to identify its meaning in the story. You may also write numbers inside a block and use brackets to group the bars together or to indicate a specific section of a bar.

If you have trouble figuring out where the numbers go in the diagram, ask yourself "Which is the big amount, the whole thing? What are the parts? How do the parts relate to each other?"

You must learn two simple but important rules:

The Whole Is the Sum of Its Parts

Bar diagrams rely on the inverse relationship between addition and subtraction: The whole is the sum of its parts. No matter how complicated the word problem, the solution begins by identifying a whole thing made up of parts.

Simplify to a Single Unknown

You cannot solve for two unknown numbers at once. You must use the facts given in your problem and manipulate the blocks in your drawing until you can connect one unknown unit (or a group of same-size units) to a number. Once you find that single unknown unit, the other quantities will fall into place.

Practice until drawing a diagram becomes almost automatic. Start with simple story problems that are easy enough to solve with a flash of insight. Figure out how you can show the relationships between quantities, translating the English of the stories into a bar model picture. Then work up to more challenging problems.

The following chapters provide a range of story puzzles from early-elementary addition to middle-school stumpers. Practice using the four problem-solving questions, and experiment with bar model diagrams.

Don't let the math scare you. Never give up.

If you stick with it, I promise, you *can* be good at math!

If you wish
to learn swimming
you have to go into the water,
and if you wish
to become a problem solver
you have to solve problems.

—George Pólya

2

Lay the Foundation: One-Step Problems

Inspired by;
Mr. Popper's Penguins ©1938 Richard and Florence Atwater

A family of four adopts several penguins and teaches them to perform tricks.

[1] During the winter, Papa read 34 books about Antarctica. Then he read 5 books about penguins. How many books did Papa read in all?

[2] When Papa opened the windows and let snow come into the living room, his children made snowballs. The girl made 18 snowballs. Her brother made 14 more than she did. How many snowballs did the boy make? How many snowballs did the children make altogether?

[3] Papa had 78 fish. The penguins ate 40 of them. How many fish did Papa have left?

[4] The family dressed in their best clothes for their meeting with the show manager. Mama had a ribbon 90 centimeters long. She had 35 cm left after making a bow for her daughter's hair. How much ribbon did Mama use to make the bow?

[5] The penguins did theater shows for 2 weeks. They performed 4 shows every week. How many shows did the penguins perform?

[6] While they were staying at the hotel, Papa put a leash on one of the penguins and took him for a walk. They climbed up 3 flights of stairs. There were 10 steps in each flight. Then the penguin flopped onto his stomach and slid down all the stairs. He pulled Papa with him all the way. How many steps did Papa fall down?

Make Your Own Math

What kind of things do the characters in your favorite story world count? Do they measure or cut things, or do crafts? What do they like to collect or share, or what are their favorite foods? What kind of math stories will you create?

[7] Write a changing-amount problem set in the world of a book or movie you enjoy. Your story will include:

- A beginning amount.
- Some type of change—joining or separation, giving or taking, growth or shrinking.
- And the final amount.

Tell any two of these numbers. Then ask your reader to figure out the third.

[8] Write a collection problem set in the world of a book or movie you enjoy. Your story will include:

- A whole amount (the collection).
- Two types or groups of people, animals, or things that are parts of the collection.

Leave either the whole or one of the parts a secret. Then ask your reader to find it out.

[9] Write a comparison problem set in the world of a book or movie you enjoy. Your story will include:

- A smaller amount.
- A larger amount.
- The difference between them.

Leave one of these numbers a secret. Then ask your reader to figure it out.

[10] Write a problem that contains some sort of this-per-that unit.

For example, where in your story world would people think about the number of legs per animal, cookies per child, dollars per item they buy, or something like that?

Your problem has three numbers:

- How many same-size units.
- The size of a single unit.
- The total amount.

Tell any two of these numbers. Then ask your reader to figure out the third.

[11] Write any kind of problem you like.

Build Modeling Skills: Multistep Problems

Inspired by:

Poor Richard ©1941 James Daugherty

A young boy grows to adulthood in colonial America.

[1: 1706] Ben helped his father make 650 tallow candles. After selling some, they had 39 candles left. How many candles did they sell?

[2: 1718] Ben sold 830 newspapers. His brother James sold 177 fewer newspapers than Ben.

(a) How many newspapers did James sell?

(b) How many newspapers did they sell altogether?

[3: 1724] Ben loved to visit the London book shops. In one shop, he found 6 shelves of books on geography, travel, and maps from around the world. Each shelf held the same number of books. There were 30 books altogether. How many books were on each shelf?

[4: 1726] Ben and his friends made a club called "The Junto" to read books and discuss ideas. Ben read 7 science books. He read 5 times as many history books as science books. How many more history books than science books did he read?

[5: 1732] Ben collected donations for many worthy organizations. He had 2,467 pounds in a bank account to start a new hospital. A friend gave him another 133 pounds. How much more money must Ben collect if he needs 3,000 pounds for the hospital?

[6: 1776–1785] While in France to negotiate a treaty, Ben went to a fancy party. There were 1,930 women at the party. There were 859 fewer men than women. How many people were at the party altogether?

Make Your Own Math

Think about the characters in your favorite story world. What do they do during the day? Do they cook, or travel, or build things? What kinds of things would they use math to figure out? Have fun making up your math stories.

[7]

Write a changing-amount problem set in the world of a book or movie you enjoy. Your story will include:

- A beginning amount.
- Some type of change—joining or separation, giving or taking, growth or shrinking.
- And the final amount.

Tell numbers for the final amount and *either* the initial amount or the change that happened. Ask your reader to find the number you leave out.

[8] Write a comparison problem set in the world of a book or movie you enjoy. Your story needs four numbers:

- ♦ A smaller amount.
- ♦ A larger amount.
- ♦ The difference between them.
- ♦ The total of both amounts.

Leave two of these numbers a secret. Then ask your reader to figure them out.

[9] Write a problem about a collection, set, or group composed of three or four parts.

Tell the total amount and the sizes of all but one of the parts.

But for some of the parts, don't give away the number. Just tell a clue—for example, how much more or less than another part. This makes a multistep puzzle-within-a-problem.

Can your reader figure it out?

[10] Write a problem that contains some sort of this-per-that unit.

For example, where in your story world would people think about the number of ounces per recipe, apples per basket, cost per book, or something like that?

Your problem has three numbers:

- How many same-size units.
- The size of a single unit.
- The total amount.

Tell any two of these numbers. Then ask your reader to figure out the third.

[11] Write any kind of problem you like.

Master the Technique: From Multiplication to Fractions

Inspired by:

The Lion, the Witch and the Wardrobe ©1949 C. S. Lewis

Four children discover a magical world with talking animals.

[1] 35 fauns came to a midnight party of fauns and dryads in the forest. There were 3 times as many dryads as fauns. How many creatures were at the party?

If the whole group split equally into 5 large circles for dancing, how many were in each circle?

[2] A studious professor had 486 books at his house, some in the library room and some in his study. There were 50 books more in the library than in the professor's study. How many books were in the study?

[3] The witch queen had 300 wolves and dwarfs as servants at her house, which was really a small castle. There were 10 more wolves than red dwarfs. The number of red dwarfs was twice the number of black dwarfs. How many black dwarfs worked at the witch's house?

[4] The beaver's wife baked a large marmalade roll for dessert. She cut $\frac{1}{6}$ of the roll for her and her husband to share, and then she sliced up $\frac{4}{6}$ of the roll for the children. What fraction of the marmalade roll was left?

[5] The beavers had a pitcher of milk. They poured $\frac{1}{2}$ of it into glasses for the children to drink with dinner. Then they poured $\frac{1}{8}$ of the pitcher into their cups of after-dinner tea. How much of the pitcher of milk did the beavers use?

[6] The witch queen's sledge got stuck in the mud and slush 24 times before she gave up and decided to walk. $\frac{2}{3}$ of those times, the witch made her prisoner get out and help push. How many times did the prisoner have to push the sledge?

[7] $\frac{2}{5}$ of the creatures waiting with the magical lion prince at his pavilion were dryads and naiads. There were 20 dryads and naiads in all. How many creatures were waiting with the prince at his pavilion?

[8] The lion prince sent 20 of the swiftest creatures to follow the wolf and rescue the prisoner. $\frac{2}{5}$ of these creatures were eagles, griffins, and other flying fighters. The rest were centaurs, leopards, and other fast-running beasts. How many of these creatures could not fly?

[9] The witch queen's evil minions used $4\frac{2}{5}$ meters of rope to bind the lion prince's legs together. They used $\frac{3}{10}$ m less of rope to tie him tightly to the stone. How many meters of rope did the horrid creatures use in all?

[10] $\frac{4}{5}$ of the sea people who sang and played music for the coronation party were mermen. The rest were mermaids. If there were 8 mermaids, how many sea people performed at the party?

Make Your Own Math

Have you had fun writing story problems of your own? Now think about how your story characters would use multiplication, division, and fractions. What sorts of things would they compare, or arrange into groups, or share between friends?

[11] Write a scaling-up problem set in the world of a book or movie you enjoy. There will be one group or part or thing that is some number of times as many or as big as something else.

Your problem has four numbers:

- The smaller amount.
- The larger amount.
- The scale factor (how many times as much).
- And the total of both amounts.

Tell any two of these numbers. Then ask your reader to figure out one (or both) of the others.

[12] Write a multistep comparison problem set in the world of a book or movie you enjoy. Your story needs four numbers:

- A smaller amount.
- A larger amount.
- The difference between them.
- The total of both amounts.

Tell the difference and the total amount. Can your reader figure out the size of the items?

[13] Write a problem about a collection, set, or group composed of three or four parts.

The numbers may include the total amount and the sizes of the parts. Or instead of telling the size of a part directly, you might use a comparison (how much more or less, or how many times as much) to make a multistep puzzle-within-a-problem.

How much information do you need to include so that your reader can figure it out?

[14] Write a problem that contains some sort of this-per-that unit. For example, where in your story world would people think about the number of cups per pint, candies per package, cost per drink, or something like that?

Your problem has three numbers:

- How many same-size units.
- The size of a single unit.
- The total amount.

Tell any two of these numbers. Then ask your reader to figure out the third.

[15] Write a problem about sharing something by dividing it. Use fractions in the problem, or let the answer be a fraction.

[16] Write a problem with a scaling-down comparison. Make one group or item some fraction of the other.

[17] Write a problem that includes some type of measurement with a mixed number.

[18] Write a problem that uses fractions with different denominators.

For an easier problem, make the denominators related to each other, like thirds and sixths. For a harder problem, use unrelated denominators.

[19] Write a story using fraction division. Tell the size of a fractional part of your quantity or group. Then ask readers to find the whole amount.

[20] Write any kind of problem you like.

5

Extend Your Skills: Measurement and Decimals

Inspired by:
The Railway Children ©1905 E. Nesbit

A mother and three children move to the countryside and meet new neighbors.

[1] When Peter was sick, Mother gave him barley water to drink. Phyllis was curious and wanted a taste also. Peter drank 0.7 liter of barley water. Phyllis drank 0.2 liter.

(a) How much barley water did they drink?

(b) How much more did Peter drink than Phyllis?

[2] The children made a bandit's lair in the attic. Peter and Bobbie captured Phyllis to hold her for ransom. Phyllis insisted on being tied up. Phyllis had a piece of string 1.85 m long. Then Bobbie found another piece of string that was 1.4 m longer than the first. Find the total length of the two pieces of string.

[3] When mother was ill, the old gentleman bought a good many gifts and packed them in a hamper to send to the children. At one shop, he bought 1 kg of grapes for £4.15 and a basket of peaches for £7.59. He gave the cashier £15. How much change did the old gentleman receive?

[4] On Bobbie's birthday, the family had fancy cranberry juice along with their tea. Bobbie, Peter, and Phyllis each drank 0.4 liter of juice. How many liters of juice did the children drink in all?

[5] Mother had £30. She wanted to take her children to a nearby town for an outing. She bought 4 train tickets. Each ticket cost £2.95. How much money did Mother have left?

[6] The Doctor had 0.9 liter of medicine at his house. He poured it into 3 bottles and gave one to the sick Russian stranger. How much medicine was in the little bottle?

[7] After the children saved the train, Mother bought flannel to make warm, new petticoats. She bought 5 meters of flannel for £8. How much did one meter of flannel cost?

[8] The Station Master kept boxes of fancy chocolate bars that passengers could buy to eat while they waited for their train. A full box of 12 bars weighed 2.34 kg. After the ceremony with the watches, the Station Master gave each of the children 4 whole bars of chocolate. Then the empty box weighed 0.06 kg. Find the weight of one chocolate bar.

[9] Mrs. Ransome posted 4 parcels. One of them weighed 1.8 kg. The other 3 parcels weighed 2.05 kg each. Find the total weight of the parcels.

[10] Peter helped the grocer carry boxes into his storeroom. One box contained 5 packets of flour. If the total weight of the flour was 5 kg 650 g, find the weight of each packet of flour.

[11] Where the landslide covered the railway, it took the workmen 7 h 30 min to clear away the debris from 6 meters of the railway track. How long did it take for the workmen to clear one meter?

[12] Bobbie and Jim told each other stories to take their minds off the darkness in the tunnel. Jim told how he'd once helped to build a stone fence. He spent 3 h 30 min every morning, carrying stones and lifting them into place. He finished the fence in 5 mornings. How much time did Jim spend working on the fence?

[13] After the old gentleman turned their house into the Three Chimneys Hospital, Mother sewed new curtains for the parlor and the 3 bedrooms. She used 3.46 m of material for each bedroom and 4.25 m of material for the parlor. How much material did mother use altogether?

[14] The usual price of a large cinnamon sticky bun was £0.60. The baker had a sale, and mother bought 4 sticky buns for Jim and the children to have with tea. Mother paid £2.20 for the buns. How much cheaper was each sticky bun during the sale?

Make Your Own Math

Where in your favorite story world would the characters use decimal fractions? Do they measure distances or weights, or do they have a decimal-based money system? What kind of math stories will you create?

[15] Write an addition story with decimal measurements.

Instead of telling the size of a part directly, you might use a comparison (how much more or less, or how many times as much) to make a multistep puzzle-within-a-problem.

How much information do you need to include so that your reader can figure it out?

[16] Write a subtraction problem with decimals, perhaps by counting money and making change. The problem may also contain addition, as a normal shopping trip might.

[17] Write a multistep comparison problem using decimals. Your story has four (or five) numbers:

- A smaller amount.
- A larger amount.
- The difference between them.
- (Optional) How many times the larger number is of the smaller number's size, or what fraction the smaller number is of the larger.
- The total of both amounts.

Tell any two of the numbers, and then ask for whichever of the missing quantities you like.

[18] Write a problem that involves multiplication with three or more decimal-fraction-sized parts. You can ask for the total product or ask for a sum or difference involving that product.

[19] Write a problem about sharing something by dividing it. Use decimals in the problem, or let the answer be a decimal.

[20] Write a problem involving time that requires the reader to convert between minutes and hours, or between seconds and minutes. Can you trick them into using a decimal conversion instead of the correct base-60?

[21] Write a problem about dividing something by measuring out mixed-decimal-fraction parts. That is, measure out parts that are greater than one but have a decimal fraction at the end (for example, 3.75). Your story has four numbers:

- The original amount (which may be a whole number or have a decimal part).
- The size of the measured-out parts (a decimal mixed number).
- The number of parts you get (a whole number).
- The leftover amount, if any.

Tell any two of the numbers, and then ask for one (or both) of the missing quantities.

[22] Write any kind of problem you like.

6

Reap the Reward: Ratios and More Fractions

Inspired by:
The Hobbit ©1937 J. R. R. Tolkien
The Lord of the Rings ©1954 J. R. R. Tolkien

A halfling joins dwarfs and a wizard on the quest to steal a dragon's treasure, and then other adventures ensue.

[1] The halfling had 3 times as many apple tarts as mince pies in his larder. If he had 24 more apple tarts than mince pies, how many of the pastries (both tarts and pies) did he have altogether?

[2] Three trolls had 123 pieces of gold. Andrew had 15 pieces of gold more than Bert. Bert had 3 pieces fewer than Charles. How many pieces of gold did Charles have?

[3] The Great Goblin had twice as many goblin soldiers as his cousin, the Gross Goblin. How many soldiers must the Great Goblin send to his cousin so that they will each have 1,200 goblin soldiers?

[4] The cave creature caught 10 small fish. He divided the fish to make 4 equal meals. How many fish did he eat at each meal?

[5] The giant bear-man baked a large loaf of whole-grain bread. He ate $\frac{1}{3}$ of the loaf himself (with plenty of honey), and he sliced $\frac{1}{2}$ of the same loaf to feed the dwarfs and their halfling friend. What fraction of the loaf was left?

[6] The king of the elves had a barrel of fine wine. His butler poured $3/4$ gallon of it into a small keg. He drank $1/2$ of the keg and gave the other half to his friend, the chief of the guards. How much wine did the elven king's butler drink?

[7] $\frac{2}{3}$ of the items in the dragon's treasure were made of gold. $\frac{1}{4}$ of the remaining part was precious gems. What fraction of the dragon's treasure was precious gems?

[8] When the king of the elves heard the dragon had been killed, he set out to claim a share of the treasure. $\frac{2}{5}$ of his army were archers. $\frac{1}{2}$ of the remainder fought with spears, and the rest carried swords. If 300 soldiers carried swords, how many elves marched out with the elven king?

[9] The master bowman gathered a small army of 600 survivors from the town the dragon had destroyed. The ratio of archers to swordsmen was 2:3. How many archers followed the master bowman to the final battle?

[10] The dwarfs rewarded their halfling friend with two chests of gold, silver, and small gems—6,000 pieces of treasure altogether. There were twice as many pieces of gold as there were gems. There were 600 more pieces of silver than gems. How much of each type of treasure did the halfling receive?

[11] The traveling wizard loaded so many fireworks into his cart that he only had time to shoot off $3/5$ of them. If he blasted 72 fireworks during the party, how many did he bring in his cart?

[12] Our halfling hero and his four friends (everyone always forgets the friend who stayed behind) enjoyed one last meal before setting off on their quest. If the hobbits shared equally all 135 sausages in the larder, how many sausages did each hobbit eat?

[13] The four halflings shared two loaves of bread for lunch on their journey. The leader wasn't very hungry, so he took a small piece and then split the rest evenly among his friends. If the other three halflings each received $\frac{3}{5}$ of a loaf, what size piece did their leader eat?

Make Your Own Math

What sort of fractions do the people in your story world use? What do they measure or build? Do they use ratios to compare things, or to keep track of ingredients in a recipe? Have fun making up your math stories.

[14] Write a scaling-up or scaling-down problem set in the world of a book or movie you enjoy.

There will be one group or part or thing that is some number of times as many or as big as something else. Or some fraction as much as the other amount.

Your problem has four numbers:

- The smaller amount.
- The larger amount.
- The scale factor.
- And the total of both amounts.

Tell any two of these numbers. Then ask your reader to figure out one (or both) of the others.

[15] Write a problem that includes some type of ratio.

The groups or parts or things are related in a more complex way. Not just "twice as much" or "one-third as big," but using a ratio like 2:3.

[16] Write a multistep problem set in the world of a book or movie you enjoy.

Have three or more parts to compare within your story. The numbers may include the total amount and the sizes of the parts. Or instead of telling the size of a part directly, you might use a comparison (how much more or less) or ratio (how many times as much) to make a multistep puzzle-within-a-problem.

How much information do you need to include so that your reader can figure it out?

[17] Write a transfer problem set in the world of a book or movie you enjoy.

Your problem needs an initial situation, and then one character gives some amount to another character, which sets up the final situation.

But be tricky about the information you give. Instead of telling the transferred amount directly, you might use a comparison or ratio to make a multistep puzzle-within-a-problem.

Go back and look at the goblin army problem for an example. How much information do you need to include so that your reader can figure it out?

[18] Write a problem about sharing something. Use mixed numbers in the problem, or let the answer be a mixed number.

[19] Write a problem that uses fractions with different denominators.

For an easier problem, make the denominators related to each other, like thirds and sixths. For a harder problem, use unrelated denominators.

[20] Write a problem that involves a fraction of a fraction.

Have three or more groups or items, where at least one is some fraction of another. Or tell a two-step problem where your character works with some fraction of a remaining part.

[21] Write any kind of problem you like.

Conquer Monster Topics: Percents, Rates, and Proportional Reasoning

Inspired by:
Tales from Shakespeare ©1807 Charles and Mary Lamb

Characters explore the heights and depths of human nature in the Bard's comedies and tragedies.

[1] The Tempest

Caliban picked sugarcane and ground it for Miranda. He brought her 2.5 kg of sugar. Miranda used 325 g of the sugar to make cookies and 1.45 kg to make cakes. How much sugar did Miranda have left? Give the answer in kilograms.

[2] Midsummer Night's Dream

When the king of the fairies sent him to find a love-potion flower, Puck flew to a certain field. There were 200 flowers in the field. Only 2 of the flowers were the type Puck sought. What percentage of the flowers were love-potions?

[3] The Winter's Tale

The infant Princess Perdita lay abandoned in a basket, along with jewelry valued at 750 gold pieces. A poor shepherd found her. He sold some of the jewels for 300 gold pieces, that he might provide a good home for his adopted daughter. What percentage of the treasure did he save to give to Perdita when she grew up?

[4] Much Ado about Nothing

When Don Pedro returned from the war, he brought a booty of 800 gold pieces. He wagered 3% of this treasure that he could trick Benedick and Beatrice into falling in love. How much money did Don Pedro wager?

[5] As You Like It

There were 60 men living with the banished duke in the forest of Arden. 55% of the men were wealthy friends and noblemen who had followed the duke. The rest were servants who came with their masters into exile. How many of the men in the forest were servants?

[6] Two Gentlemen of Verona

Faithless Proteus sent his page (fair Julia in disguise) to buy a fancy gift for the lady Sylvia, even though he knew Silvia did not love him. The gift cost 1,350 Lira plus a 3% tax. How much did Julia pay for this unwelcome gift?

[7] The Merchant of Venice

The merchant Antonio had three ships at sea. If the average profit from a merchant ship's journey was 2,500 ducats, how much did Antonio expect to gain when his ships came to dock?

[8] Cymbeline

After wicked Iachimo stole her bracelet, Imogen got lost in the forest. She stopped to drink at a small spring, where water flowed out between the rocks before running off to join a stream. 15 liters of water fell from the spring every 6 minutes. Find the rate of flow in liters per minute.

[9] King Lear

After being abandoned by his wicked daughters, King Lear wandered through the countryside. It took him 9 days to travel 45 km. At this rate, how long would it take him to travel 125 km to the earl of Kent's castle at Dover?

[10] Macbeth

The three witches heated their cauldron slowly, so the potion came almost to a boil. The potion produced 12 bubbles in 48 Seconds. At this rate, how many bubbles would it make in one minute?

[11] All's Well That Ends Well

After Count Bertram rejected her as his wife, Helena traveled 750 kilometers to Florence, Italy. If her average rate of travel was 30 km per day, how long did the journey take?

[12] Twelfth Night; Or, What You Will

Viola, disguised as the page Cesario, visited the lady Olivia to return her ring. They had tea in the garden, which had the shape of a rectangle 60 m long by 40 m wide. Find the ratio of the length to the width to the perimeter of the garden.

[13] Hamlet, Prince of Denmark

After a late-night game of cards, the ratio of Horatio's winnings to Hamlet's money was 2:5. Horatio won 24 guilders. How much money did Hamlet have?

[14] Othello

The general Othello led an army of 14,000 men to defend Cyprus against the invading Turks. 15% of the army were gentleman officers, and the rest were common soldiers. How many gentleman officers were under Othello's command?

[15] Pericles, Prince of Tyre

The lady Marina, captured by pirates and sold as a slave, earned money by teaching. Her master allowed her to keep 4% of her earnings to spend on herself. If Marina kept 6 talents, how much money did she earn?

[16] Romeo and Juliet

Romeo put on a mask and went uninvited to a party thrown by Lord Capulet. The number of men at the party was $5/8$ the number of women. If there were 24 more women than men, how many people attended Lord Capulet's party?

Make Your Own Math

Where in your favorite story world would the characters use rates, ratios or percents? Do they measure speed, or buy steampunk gears by the pound, or pay a worker by the hour? When might they need to compare a part to a whole thing? What stories will you create?

[17] Write a scaling-up or scaling-down problem using a ratio, set in the world of a book or movie you enjoy.

Your problem has five numbers:

- The smaller amount, and the larger amount.
- The difference between them, and the total of both amounts.
- The ratio comparison.

Tell the ratio and any *one* of the other numbers. (You may tell more than one number, if you wish.) Then ask your reader to figure out whichever of the remaining numbers you like.

[18] Write a problem that uses percents.

Where in your story world would a character work with percent calculations? Think about comparing parts to a whole group, making a wager, dividing the spoils from a raid, and things like that.

The numbers in a percent problem may include whatever counts as the *base* (100%), the part of that thing we are interested in, and the percentage value. Or you might use the *complement*, which is the percentage value for the part we are *not* interested in.

Tell two of the percentage (or the complement), the part, and the base. Then ask your reader to figure out the missing number.

[19] Write a multistep problem set in the world of a book or movie you enjoy. Have two or more parts to compare within your story.

The numbers may include the total amount and the sizes of the parts. But instead of telling the size of a part directly, use a fraction, ratio, or percent comparison to make a multistep puzzle-within-a-problem.

How much information do you need to include so that your reader can figure it out?

[20] Write a rate problem, where your characters do something or go somewhere at a this-per-that rate, or they have some amount of this or of that which must be processed or poured out or filled up or traveled.

Your problem has three numbers:

- How fast your characters work or travel (the rate).
- How much they get done.
- How long it takes them.

Tell any two of these numbers. Then ask your reader to figure out the third.

[21] Write any kind of problem you like.

8

Move Toward Algebra: Challenge Problems

Inspired by:
The Lord of the Rings ©1954 J. R. R. Tolkien
Star Wars ©1977 George Lucas, 20th Century Fox
Dealing with Dragons ©1990 Patricia C. Wrede
Idylls of the King ©1859–1885 Alfred, Lord Tennyson

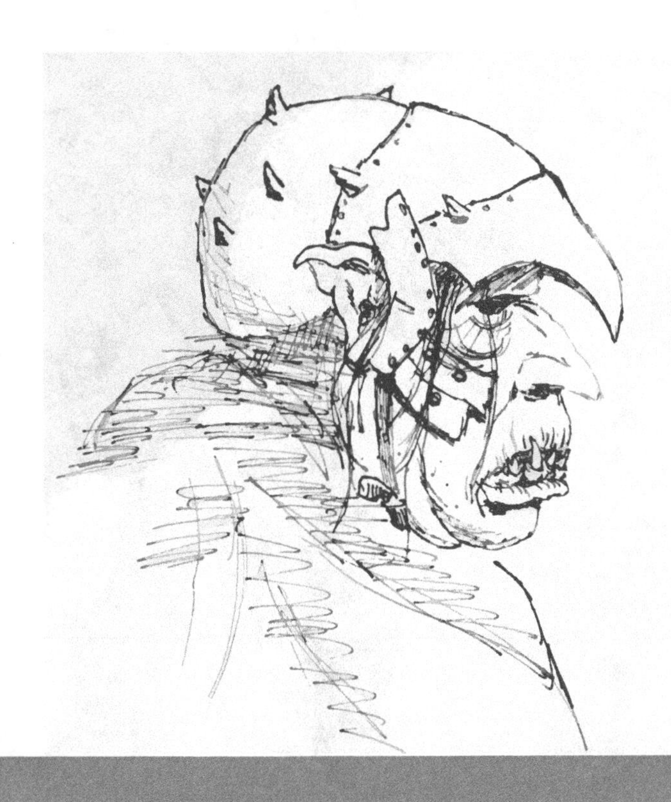

[1] The day before the climactic battle at the gate, a company of 450 orcs camped among the necromancer's great host. But an argument broke out over dinner, and $\frac{1}{3}$ of them were killed. Then $\frac{2}{5}$ of the remainder died when a drunken troll stumbled through their camp during the night. How many of the orcs survived to join the morning's battle?

[2] The necromancer's tailor has 32 yards of black cloth to make traveling capes for his ghostly servants. The tailor needs $3\frac{3}{4}$ yards for each cape. How many capes can he make? How much additional cloth will he need to make all 9 capes?

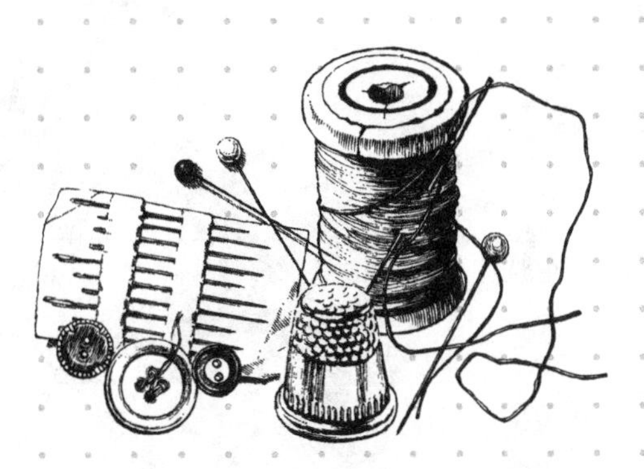

[3] The rogue space smuggler Peyton and her first mate Spider performed maintenance work on their spaceship. Peyton spent $\frac{3}{5}$ of her money upgrading the hyperspace motivator. She spent $\frac{3}{4}$ of the remainder to install a new blaster cannon. If Peyton spent 450 money chips altogether, how much money did she have left?

[4] A captive princess worked all afternoon cleaning and organizing part of Dragon Philip's treasure. $\frac{1}{4}$ of the items she sorted were jewelry. 60% of the remainder were potions, and the rest were magic swords. If there were 48 magic swords, how many pieces of Dragon Philip's treasure did she sort in all?

[5] Knowing the goblin horde would soon be upon them, Queen Emily placed 1,500 guards atop the outer wall at the canyon entrance that protected her fortress. 300 of the queen's guards were swordsmen and the rest were archers. What percentage fewer swordsmen were there than archers?

[6] King Arthur gathered several of his knights to send them on a quest. He had a bag of silver coins to give them for provisions along the way. If he gave each knight 8 coins, he would have 4 coins left in his bag. But if he gave only 5 coins to each knight, he would have 40 coins left. How many knights were going on the quest?

[7] King Arthur had a small chest of jewels in his treasury, gifts from the many foreign dignitaries who visited his court. $\frac{1}{3}$ of the stones were rubies, $\frac{1}{9}$ were emeralds, and $\frac{1}{5}$ of the remainder were diamonds. If there were 25 diamonds, how many jewels did King Arthur have in the chest?

[8] King Arthur and Merlin went shopping at the Camelot Fair. They both had the same number of coins. Then Arthur spent 50 coins on new tapestries for the walls in his castle, to keep down the winter chill. And Merlin spent ⅓ of his money on new books and scrolls to study.

When they compared their purses at the end of the day, the ratio of King Arthur's money to Merlin's was 5:4. How many coins did they each have at the beginning of the day?

[9a] Captain Henry wants to upgrade the old hyperdrive configuration on his starship. The power controls consist of two toggle switches on his dashboard, and each switch can point either up or down.

Every arrangement of these switches signals the engine to run at a different power level. With the current configuration, the captain can apply four levels of hyperdrive power:

- down, down (the lowest level)
- down, up
- up, down
- up, up (the highest level)

But his engine can actually run at ten different power levels. How many additional switches should the ship's engineer install to let Captain Henry signal all ten hyperdrive levels?

[9b] Chief Engineer John fixes the hyperdrive controller, putting in the new switch(es) that the captain ordered.

Then John decides it's time to upgrade the ship's engine, too. But he doesn't want to make the controls too confusing.

How many additional hyperdrive power levels will the new toggle-switch configuration handle, without installing still more switches?

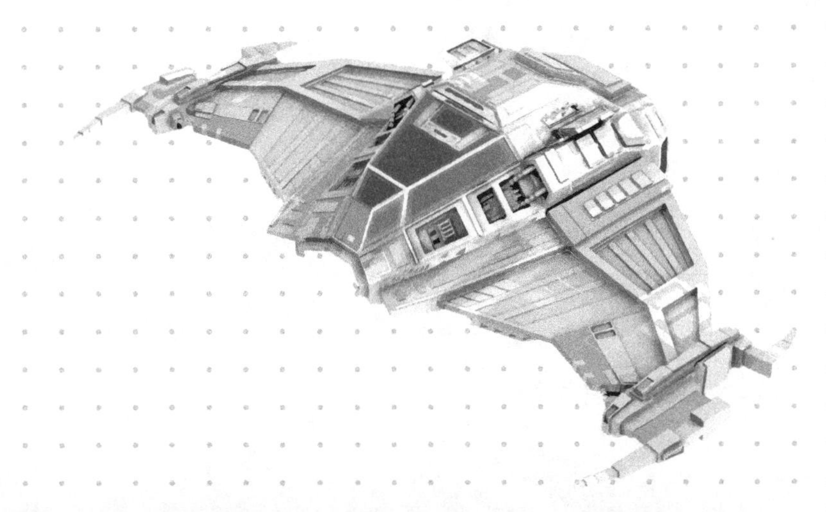

Make Your Own Math

How do the people in your story world use math? Do they study it in school? What kind of numbers do they use, and do they know how to write equations? Have fun playing around with math stories.

[10] Write a multistep problem set in the world of a book or movie you enjoy. Have three or more parts to compare within your story.

The numbers may include the total amount and the sizes of the parts. Or instead of telling the size of a part directly, you might use a comparison (especially a fraction or ratio) to make a multistep puzzle-within-a-problem.

How much information do you need to include so that your reader can figure it out?

[11] Write a fraction-of-the-remainder problem set in the world of a book or movie you enjoy.

Tell a two-or-more-step problem that begins with an initial condition, then something changed, and there's a part left over, so your character works with some fraction of that remaining part.

What questions will you ask? How much information do you need to include so that your reader can figure it out?

[12] Write a problem that uses percents. Where in your story world would a character work with percent calculations? Think about sharing out a treasure, shopping with sales or discounts, paying taxes, saving or loaning money, and things like that.

The numbers in a percent problem may include whatever counts as the base (100%), the part of that thing we are interested in, and the percentage value. Or you might use the complement, which is the percentage value for the part we are *not* interested in.

Tell two of the percentage (or the complement), the part, and the base. Then ask your reader to figure out the missing number.

[13] Write a problem that uses both fractions and percents.

When might your character need to find a percentage of a fraction, or a fraction of a percentage? What kind of situation would lead to such a calculation?

[14] Write a percent comparison problem.

Why might your character be comparing two (or more) things and notice that one of them is some percentage more or less or greater or fewer than the other? What questions can you ask about such a situation?

[15] Write any kind of problem you like.

Create Your Own Word Problems

What do you get when you cross a library book or favorite movie with a math worksheet?
A great alternative to math homework!

Over time
you will get
the wrong answer
more times
than you get
the right answer.
That's not a problem!
You've learnt
what doesn't work,
just try again.

—Sophie Carr

The Story Problem Challenge

THE RULES ARE SIMPLE:

(1) Choose a worksheet calculation to be the basis for your word problem.

(2) Solve the calculation.

(3) Consider where these numbers could make sense in your book or movie universe. How might the characters use math? What sort of things would they count or measure?

Do they use money? Do they build things, or cook meals, or make crafts? Do they need to keep track of how far they have traveled? Or how long it takes to get there?

(4) Write your story problem.

To make the game easier, you may change the numbers to make a more realistic problem. But you must keep the same type of calculation. For example, if your worksheet problem was 18 ÷ 3, you could change it to 18 ÷ 6 or 24 ÷ 3 or even 119 ÷ 17 to fit your story, but you can't make it something like 18 – 3.

Remember that some quantities are discrete and countable, such as hobbits and fireworks. Other quantities are continuous, such as a barrel of wine or a length of fabric. Be sure to consider both types when you are deciding what to use in your problem.

Then share your problem with friends, and you try their problems. Can you stump each other?

Problem Situations for Addition and Subtraction

People often think of addition and subtraction in stories that involve movement: joining or leaving, giving or taking, growth or shrinkage. But you can also write stories about comparison —how much larger or smaller (lighter or heavier, more or less expensive, etc.) is this thing than the other one. Or you can use classification, sorting a collection of people or items into different categories.

Here are some story problem situations you might try:

- The story gives the parts. Find the whole amount.
- The story gives the whole and one (or more) of the parts. Find the missing part.
- The story tells one of the parts and the difference between that and the other part. Find the missing part and/or the whole amount.
- The story tells the whole and one part. Find the difference between the parts.
- The story tells the whole amount and the difference between the parts. Find the parts.

Problem Situations for Multiplication and Division

People often think of multiplication and division as counting or splitting groups of items. But you can also think of things that grow or shrink, so their new size is some number of times as big or some fraction the size as before.

Here are some story problem situations you might try:

- The story says that one part is some number of times greater than the other part, and it gives either the whole amount or the size of one of the parts. Find the missing information.
- The story has a whole thing made of some number of equal parts. Find the missing piece of information: the whole amount, the size of the part, or the number of parts.
- In the story, one part is a given multiple of the other part, and the story tells the whole amount. Find either part, or both of them, or the difference between them.
- In the story, one part is a given multiple of the other part, and the story tells the difference between the parts. Find either part, or both of them, or the whole amount.

Problem Situations with Fractions and Decimals

If you are working with fractions, think about sharing something continuous, like pizza or fabric. For decimals, you'll probably want stories about measurements or about money.

Here are some story problem situations you might try:

- In the story, one part is some fraction (or decimal) of the whole amount. Two of the quantities are given: the fraction (or decimal) and either the part or the whole. Find the missing quantity.
- In the story, one part is some fraction (or decimal) of another part. Two of the quantities are given: the fraction (or decimal) and one of the parts. Find the other part, the whole amount, and/or the difference between the parts.

- In the story, one part is some fraction (or decimal) of another part. The story tells the fraction (or decimal) and the whole amount. Find the two parts.

Problem Situations with Ratios and Rates

Ratios compare two or more objects in proportion, like the amounts of butter, flour, and sugar in a recipe for scones. Rates are a this-per-that ratio, like sausages per hobbit or yards of cloth per cape.

Here are some story problem situations you might try:

- The story tells the ratio relationship between two (or more) parts and also gives one part or the whole amount. Find the missing parts and/or the total amount.
- The story tells the ratio relationship between two parts and also gives one part or the difference between the parts. Find the missing part and/or the total amount.
- The story tells the ratio relationship between two parts and also gives either the whole amount or the difference between the parts. Find the size of the parts.

Problem Situations with Percents

Where in your story world would a character work with percent calculations? Think about sharing out a treasure, shopping with sales or discounts, paying taxes, saving or loaning money, and things like that.

Percents are one of the math monsters, the toughest topics of arithmetic. The most important step in solving any percent problem is to figure out what quantity is being treated as the *base*, the whole thing that equals 100%. Everything else in the problem is measured in comparison to the base.

Here are some story problem situations you might try:

- In the story, one part is some percentage of the whole amount (the base). Two of the quantities are given: the percentage, the part, and/or the base. Find the missing quantity.
- In the story, one part is some percentage of the whole amount (the base). Two of the quantities are given: the percentage, the part, and/or the base. Find the *other* part of the whole.
- In the story, one part is some percentage of another part (the base). The whole amount is *not* the base. Two of the quantities are given: the percentage and either part. Find the other part, the whole amount, and/or the difference between the parts.
- In the story, one part is some percentage of another part (the base). The whole amount is *not* the base. The story tells the percentage and the whole amount. Find the two parts.
- In a percent comparison story, one amount is some percentage greater or less than another amount (the base). Two of the quantities are given: the smaller amount, the larger amount, the difference between them, and the percentage. Find either (or both) of the other two.

A Note about Copyright and Trademarks

Old books are in the public domain, so you can always use characters like Robin Hood, Sherlock Holmes, or Winnie-the-Pooh (but not the newer Disney version with the red jacket). But most books and movies are the protected intellectual property of their authors or estates, or of the company which bought those rights.

When you write problems for your own private use, feel free to use your favorite characters from any story. That's like fan fiction, secret, just for your own pleasure.

But if you decide to share your creation beyond your own home or classroom, then be sure to "genericize" it first. Change or remove the proper names, using general descriptions instead.

For example, if you love the Harry Potter series, you might want to use Harry or Hermione in your story problems. Instead, write about "the boy wizard destined to fight an evil sorcerer." Or "the bright young witch who can master any spell."

Or if you like the Star Wars movies, you might write about "an interstellar justice warrior with an energy sword." Or "an alien master of martial arts training a cocky but inexperienced apprentice."

We'd love to add your story to the Student Math Makers Gallery.
tabletopacademy.net/math-makers

...and may the Math be with you!

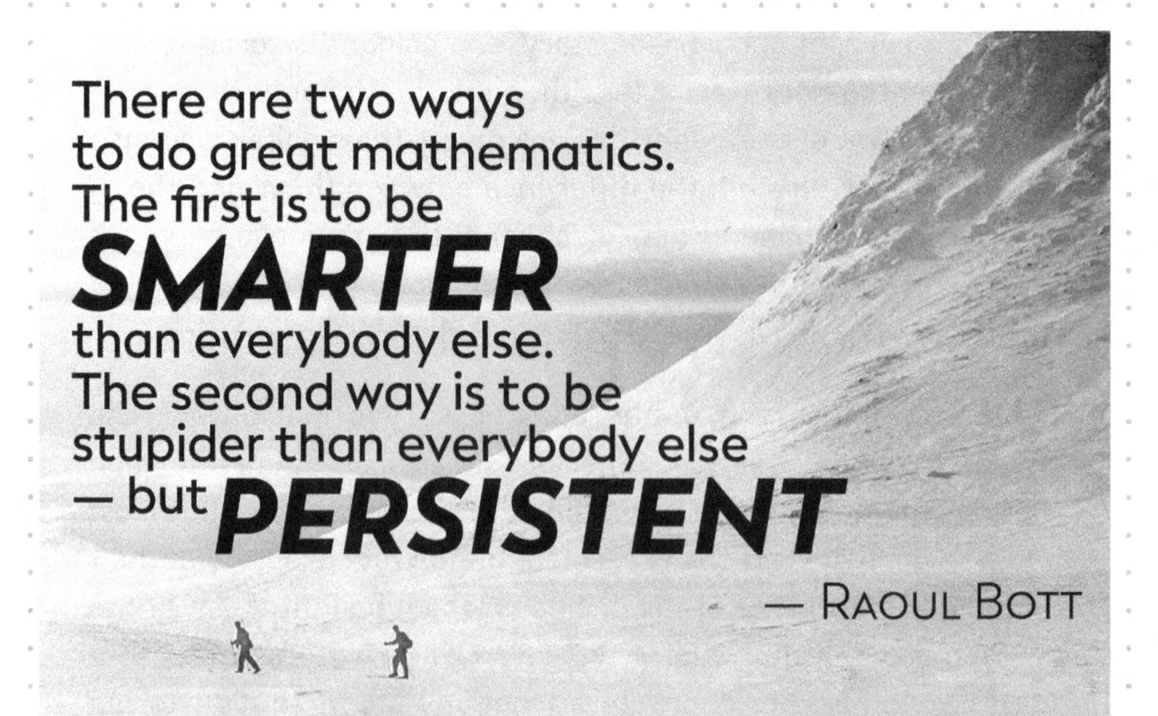

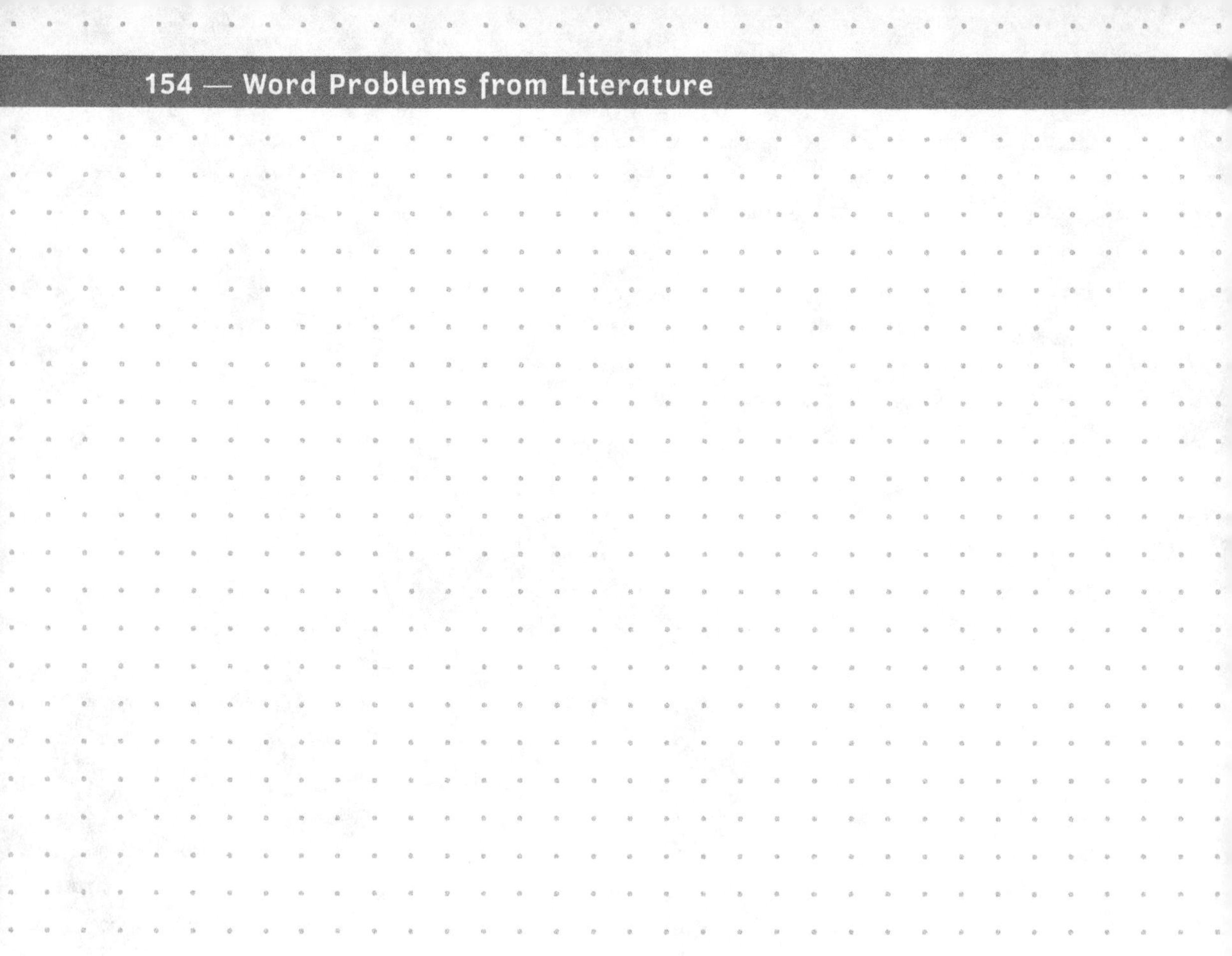

About the Author

FOR MORE THAN THREE DECADES, Denise Gaskins has helped countless families conquer their fear of math through play. As a math coach and veteran homeschooling mother of five, Denise has taught or tutored every level from preschool to precalculus. She shares math inspirations, tips, activities, and games on her blog at *DeniseGaskins.com.*

Denise encourages everyone to look at math with fresh eyes. "We want to explore the adventure of learning math as mental play, the essence of creative problem solving. Mathematics is not just rules and rote memory. Math is a game, playing with ideas."

Credits

"Storying—encountering the world and understanding it..." Gordon Wells, quoted in *Writing in the Teaching and Learning of Mathematics* by J. Meier and T. Rishel.

"Math ain't about numbers..." Cliff Stoll, "Klein Bottles," posted by Numberphile on YouTube, June 22, 2015.

"If you wish to learn swimming ..." George Pólya, *Mathematical Discovery: On Understanding, Learning and Teaching Problem Solving,* Wiley, combined edition 1981.

"Over time you will get the wrong answer..." Sophie Carr in "Maths is all around: Increasing the probability that more girls will pursue mathematics related careers," Womanthology website, April 18, 2018.

"There are two ways..." Raoul Bott, quoted in The MacTutor History of Mathematics online archive.

Cover art by SergeyNivens/DepositPhoto. Illustration copyrights belong to various artists, licensed from Depositphotos.com. Shakespeare illustrations by Elizabeth Shippen Green Elliott. Author photo by Christina Vernon.

CPSIA information can be obtained
at www.ICGtesting.com
Printed in the USA
BVHW022029230223
659102BV00009B/201